Leonardo da Vinci (1452-1519) was born in Italy, the son of a gentleman of Florence. He made significant contributions to many different disciplines, including anatomy, botany, geology, astronomy, architecture, paleontology, and cartography.

He is one of the greatest and most influential painters of all time, creating masterpieces such as the *Mona Lisa* and *The Last Supper*. And his imagination led him to create designs for things such as an armored car, scuba gear, a parachute, a revolving bridge, and flying machines. Many of these ideas were so far ahead of their time that they weren't built until centuries later.

He is the original "Renaissance Man" whose genius extended to all five areas of today's STEAM curriculum: Science, Technology, Engineering, the Arts, and Mathematics.

You can find more information on Leonardo da Vinci in *Who Was Leonardo da Vinci?* by Roberta Edwards (Grosset & Dunlap, 2005), *Magic Tree House Fact Tracker: Leonardo da Vinci* by Mary Pope Osborne and Natalie Pope Bryce (Random House, 2009), and *Leonardo da Vinci for Kids: His Life and Ideas* by Janis Herbert (Chicago Review Press, 1998).

Little Leonardo's™ MakerLab ROBOTS

Written by
BART KING

Illustrated by
GREG PAPROCKI

GIBBS SMITH
TO ENRICH AND INSPIRE HUMANKIND

To the least robotic person I know:
Evan King.

Manufactured in China in January 2019 by
Crash Paper Co.

First Edition
23 22 21 20 19   5 4 3 2 1

Text © 2019 Bart King
Illustrations © 2019 Greg Paprocki

All rights reserved. No part of this book may be reproduced by any means whatsoever without written permission from the publisher, except brief portions quoted for purpose of review.

Published by
Gibbs Smith
P.O. Box 667
Layton, Utah 84041

1.800.835.4993 orders
www.gibbs-smith.com

Designed by Sky Hatter and Renee Bond

Gibbs Smith books are printed on either recycled, 100% post-consumer waste, FSC-certified papers or on paper produced from sustainable PEFC-certified forest/controlled wood source. Learn more at www.pefc.org.

Library of Congress Control Number: 2018951142
ISBN: 978-1-4236-5116-1

INTRODUCTION

Here is something that everyone knows: *Robots are cool!* But what *is* a robot, exactly? Robots come in all different shapes and do all sorts of different jobs. A robot could be a machine that mows a lawn or cleans the pool. A robot might be a **drone,** flying high in the sky. Some robots can dive into the ocean, or even volcanoes. Others can go into outer space. Right now, there is even a robot with wheels on Mars.

Here are some basic robot types:

> **Programmed robots** do one job over and over.
>
> **Mobile robots** can move around.
>
> **Remote-controlled robots** are run by humans.
>
> **Independent robots** can move and think on their own.
>
> **Androids** are robots that look and act like humans.

A robot does not need to look like a person. Even so, a robot can be *like* a person. A robot is made of lots of different parts, like you are. It even has a computer inside to help it think. That computer is like your brain.

But there are also simple robots that are not really like you. An example might be as simple as a windup toy. Or we could make a robot out of old cans and pipe cleaners.

When using this book, remember:

- Read through an activity before beginning.
- Gather the supplies you'll need.
- If the activity doesn't work perfectly the first time, try again.
- Always wash your hands when you're done with each activity.
- Have fun.

DREAM ROBOT

What You Need:

- Paper
- A pencil
- Crayons or colored markers
- Your imagination

What You Do:

1. Think of a problem. This might be a big problem, like cleaning up all the trash in the oceans. Or maybe it's finding dogs that get lost from their owners. The problem could even be too many bread crumbs in your toaster at home.

2. Now imagine a robot that could help solve this problem. What would this problem-solving robot need to do? Don't worry about having to actually *make* this robot. For right now we're just imagining.

3. Sketch a picture of your robot in pencil. As you draw, think about your robot's challenges. For example, if it's a robot that makes beds, will it be too short to tuck the sheets in? Or will the robot be tall, so it can bend over the whole bed and spread the blanket?

4. Is your robot tough looking? Cute? Does it look like a human? Or maybe a kazoo? Will it have wheels? Legs? Something else?

5. After sketching your robot, use your colored markers to make it look good.

6. Finally, give your robot a cool name. If you're stuck, try adding *-bot, -oid, -ator,* or *-tron* to its job title. For example, a dog-walking robot could be the *Leash-bot*. The word *Prime* can be added to any robot name, along with some random numbers. So now you might have the *Leash-bot Prime 3000*.

TRY THIS: Having trouble getting an idea for a robot? Take a nap. That's what a 15-year-old boy named Walter Lines did. In 1897, Walter woke up with an idea for a gizmo he could zoom around on. He called it a *scooter*! Or you can make a **decision flowchart** to help you decide what kind of robot to create. Here's an example of a simple decision flowchart. Answer the yes-or-no question in the diamond-shaped box at the top, then follow the arrow to the next box your answer points to.

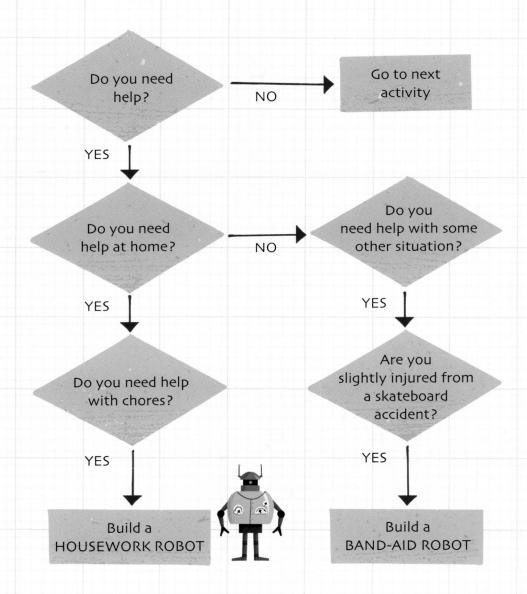

5

BE A TINKER

Someone who **tinkers** likes to figure out how things are put together. You can start being a tinker by taking stuff *apart*.

What You Need:

- ⊠ Your parents or another adult to help
- ⊠ Old stuff, gizmos, gadgets, junk, or trash
- ⊠ Tools
- ⊠ Goggles for eye protection
- ⊠ Gloves

What You Do:

1. Ask your parents if there are any gadgets around that you can take apart.

2. Don't have any good junk around your house? Take a trip to a dollar or thrift store. Look for pull-back toy cars (ones where you pull them backward and they roll forward). Also good are battery-operated fans, balls, and spinning tops with lights.

3. If your gadget has a power cord, make sure it's unplugged (and see the Safety Note at the top of page 7). If nobody will ever use this gadget again, cut the cord off to be safe.

4. Look at your gadget. Can you figure out how it was put together? Put on your goggles and gloves, then take it apart by working backward from how it was put together. Is there a screw somewhere? If so, unscrew it. A tab or button? Press it to see what happens.

5. *No hammering.* If you think you have to bash something to take it apart, just leave it alone. Bashing isn't tinkering.

6. Do you have to pry or push to take things apart? If so, pry *away* from you.

7. A tinker also tries to put gadgets back together, so be sure to save every piece. Take pictures as you take it apart. That way you won't forget a piece.

SAFETY NOTE: If your gadget uses batteries or electricity, you might find a **capacitor** inside. A capacitor looks like a little battery with legs. If you see a capacitor, don't touch it. It might shock you. Instead, ask an adult to follow these instructions to discharge the capacitor:

1. Wearing gloves and using a plastic- or wood-handled screwdriver, touch the metal end of the screwdriver to both capacitor legs at the same time.
2. You may see sparks, which indicate that the capacitor is charged.
3. Keep touching the screwdriver to the legs until the sparks disappear.
4. Now the capacitor is safe to touch.

WHERE TO FIND THINGS TO TINKER WITH

Robots are fun because you can find parts to build them *everywhere*. A colored marker cap can be a robot nose. You can take apart action figures for arms, legs, and heads. Wire from a spiral notebook is great for decorations and antennae. Here are other good places to find robot parts:

Old Toys
Cars
Dolls and action figures
Stuffed animals
Bathtub toys
Dog chew toys
Old Play-Doh containers
Construction sets (Lego, Erector, etc.)

Household and Junk Drawer Stuff
Hand fans
Aquarium pumps
Key chain lights
Old CDs and DVDs
Chopsticks
Straws
Thread spools
Paper clips
Buttons
Brass fasteners

Recyclables
Toilet paper and paper towel tubes
Plastic or paper cups
Soda bottles
Breath mint tins
Soda and soup cans
Pie pans
Paper and cardboard
Bubble wrap
Foil wrap

Tinker Leftovers
Gears
Switches
Magnets
Belts
Screws

Weird Stuff
Badminton shuttlecocks

7

MAKE A ROBOT HEAD

What You Need:

- ⌑ An old tennis ball
- ⌑ A marker
- ⌑ An adult to help
- ⌑ An X-Acto knife
- ⌑ Gloves
- ⌑ A pair of googly eyes (or art supplies to make your own)
- ⌑ Other things for decoration

What You Do:

1. Draw a line on the ball with a marker where a mouth will be cut.

2. Ask the adult to put on gloves. Then have them carefully cut through the ball where the line is. (Make sure they cut away from themself.)

3. Pinch in the sides of your robot head on either side of the mouth and it will open. Stick something in the mouth for fun, like an eraser or a dog toy.

4. Attach the googly eyes. You can also add hair, ears, antennae, or other touches.

5. Save this head for some of the robot projects coming up. You can attach the head to a robot body with things like a suction cup, glue, wire, or tape.

TRY THIS: Another fun idea to use for a robot head is a toilet paper roll. These are easy to decorate and to top off with hair.

MAKE A SCAVENGER ROBOT

If you **scavenge,** you are using things that no one else wants. For this activity go through junk and even trash to find cool robot parts (see Where to Find Things to Tinker With on page 7).

What You Need:

NOTE: This list only includes *some* of the materials you could use. Use your imagination to find other things.

- Cardboard and other building materials
- Foil wrap
- Buttons
- Clear tape
- Double-sided mounting tape
- Scissors
- Ruler or tape measure
- White glue
- Pencil
- Screwdriver
- Hole punch
- Rubber bands
- String
- Brass fasteners

What You Do:

1. Remember the drawing you made in the Dream Robot activity on page 4? See if you can make it. Or design a different robot and build that. Don't worry. This robot doesn't have to work. Just try to make a robot that looks like it *could* work.

2. If you're not sure where to begin, start with the head you made from a tennis ball in the Make a Robot Head activity on page 8.

3. Cardboard is a great material for robot parts. Have an adult help you cut out cardboard pieces that match your design.

4. Wrap robot parts in foil to give your robot a metal look.

5. Pizza box cardboard and buttons make a great robot console.

What You Need to Know:

It's hard work making a robot. For example, it took almost 30 years to make the walking robot called ASIMO. So take your time and be patient.

CANDY-TRON PRIME 7000

What You Need:

NOTE: This list only includes *some* of the materials you could use. Use your imagination to find other things.

- Plastic candy or breath mint containers (like Tic Tac containers, some with and some without the candy still in them)
- Thin plastic tubes (like Pez dispensers, without their heads)
- Plastic spice containers
- Small snap lids to spice containers and other storage containers
- Clear tape
- Double-sided mounting tape
- Model glue or white glue
- Pen or marker caps

What You Do:

1. Stand up two of the thin plastic tubes next to each other on a flat surface to use for legs. Put mounting tape or glue on their tops.

2. Stick a plastic container or lid to the mounting tape or glue on the tops of the two legs. That's the robot's body.

3. Use two more of the thin plastic tubes for arms. Glue or tape marker caps to one end of each of the plastic arm tubes.

4. Glue or tape the arms to the body.

5. Glue or tape a Tic Tac container with some candy in it on top of the body. That's the head.

6. Put a spice container lid on top of the head. Pop its lid up!

7. Decorate your robot by gluing or taping pieces of candy to it. These can be used for a mouth, nose, eyes, ears, buttons, and other things.

8. If your robot feels like it wants to fall over, it may be top heavy. Try gluing or taping coins to its legs. This will help its balance.

PULLEY POWER

Machines can help us (and robots!) do work. For example, a **pulley** is a machine that makes it easier to pull things. Many robots have pulleys in them. This activity demonstrates how a pulley works.

What You Need:

- Two assistants (your age or older)
- One rope, like a jump rope
- Two brooms, mops, broom handles, or other long sticks

What You Do:

1. Have your assistants stand facing each other. They should be far enough apart that you can stand between them with your arms outstretched.

2. Give each assistant a broom to hold in front of them at waist height, facing from side to side, not up and down.

3. Grab each of the two brooms, one with each of your hands, and try to pull them both toward you.

4. Your assistants should hold on tight and try not to let you pull the brooms toward you. As strong as you may be, you probably won't be able to pull the brooms toward you.

5. With your assistants standing in the same positions they were before, holding their brooms horizontally, tie one end of the rope near the end of one of the brooms.

6. Loop the rope tightly around the two brooms three times.

7. Holding onto the free end of the rope, stand *behind* one of the assistants and pull on the rope. This time you can probably pull the brooms together. You're using **mechanical power.**

CUP-BOT

What You Need:

- A paper or Styrofoam cup
- A small windup toy, pull-back toy car, or remote-controlled toy car
- Materials for decoration, like markers, stickers, and googly eyes
- Scissors
- Tissues or newspaper

What You Do:

1. Look at the picture below. This should give you an idea of what you are aiming for.
2. To get started, decorate your cup with your own designs and stickers.
3. Add googly eyes and antennae.
4. Place the cup over the toy car or windup toy. It might fit perfectly. If so, you rock!
5. If the car or toy is too *big,* cut pieces out of the sides of your cup to make room.
6. If the car or toy is too *small,* stuff enough tissues or newspaper around the inside edge of the cup to hold it in place.
7. If you're using a remote-controlled car, turn it on and watch the robot move around. If it's a windup toy or pull-back car, either lift up the robot and wind it or pull it back, then set it down and watch it move.

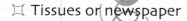

BOAT-BOT

Robots need **power** to move. That power can come from electricity, magnets, or other sources. This robot uses the power of **gravity** acting on water.

What You Need:

- ☐ 2 paper, plastic, or Styrofoam plates (not too big)
- ☐ White glue
- ☐ A paper, plastic, or Styrofoam cup
- ☐ A pen or small screwdriver
- ☐ A bendable drinking straw (the longer the better)
- ☐ A bathtub, outdoor kiddie pool, or big sink

What You Do:

1. Spread glue around the rim of one plate. Place the second plate on the first one upside down. The two plates should make a flying saucer shape.

2. With a pen or small screwdriver, carefully poke a hole in the side of the cup near the bottom. It should be just big enough to fit the straw into.

3. Slide the drinking end of the straw into the hole. Spread glue around the hole so it's watertight.

4. Glue the bottom of the cup to the bottom of one of the two plates.

5. **IMPORTANT:** The end of the straw needs to reach below water level when the Boat-Bot is floating. If the straw's not long enough to do that, try using a longer straw, or gluing or taping two straws together.

6. Give your Boat-Bot some time to let all that glue dry. Try setting a light book on top of the cup. Wait until the next day to test the Boat-Bot in water.

7. When the glue has dried, fill your tub or pool with water. Set your Boat-Bot in the water with the cup on top. Make sure the end of the straw is underwater.

8. Now fill the cup with water. Gravity is pushing the water out of the cup, through the straw, and into your tub or pool . . . which moves the boat.

GRAVITY-BOT

This robot also uses the power of gravity . . . but in a different way than the Boat-Bot on page 13.

What You Need:

- Two 3 x 5 inch index cards
- A small cardboard box about the same size as the index cards
- Tape
- String
- Foil
- Hole punch
- Tape measure
- Decorations to use for the robot's head, arms, feet, antenna (straws, ribbon, etc.)

What You Do:

1. Fold one index card so that there's a tent shape in the middle, like in the top picture below.

2. Place the folded card on top of the other index card and tape it there.

3. Fold the ends of the two cards, like in the bottom picture below, so that about ½ inch of the bottom card and a small bit of the top card are folded up.

4. Punch a hole through the taped index cards in the center of each of the two ends you folded up. Make sure the holes go through both the top and bottom cards.

5. Thread about 3 feet of string through both of those holes and *over* the tent shape in the top card, as shown in the bottom picture below. Set these index cards aside.

6. Cover the cardboard box with foil to give it a metal look. This is your robot's body.

7. Add decorations to the box for the robot's head, arms, legs, and any other cool stuff you'd like.

8. Tape the index cards to the back of your robot's body so that the tent shape faces out and the two holes are at the top and bottom.

9. With your completed Gravity-Bot laying flat on a surface, grab the string just above the robot's head with one hand. Hold the other end of the string in your other hand, about a foot below the bottom of the robot.

10. **IMPORTANT:** Stretch the string as tightly as you can by moving your top and bottom hands apart. You'll notice that the string will push hard against the tent part of the index card when you do this.

11. Lift the robot up, holding onto the string with both hands so that the robot's head is on top and its legs are on the bottom. Keep the string tight while you do this.

12. As you hold up the robot, move your two hands a little closer together. That will loosen the tension on the string. The robot should start sliding down the string. Now pull your hands farther apart again, making the string tight. The robot should stop!

13. To start over, move the robot back to the top of the string, pull the string tight, and hold it up again.

14. Turn on some music and see if you can move your robot in time to the music.

What You Need to Know:

When one thing rubs against another, that's **friction**. For example, bicycle brakes usually use the power of friction. If you squeeze a bike's brake handles, the brakes rub against the wheels, slowing them down. When you pull the Gravity-Bot's string tight, friction slows it down. But loosen your hold on the string, and gravity takes over and moves the robot down.

FLAT ROBOT

What You Need:

☐ An empty cereal box

☐ Pencil

☐ Ruler

☐ Scissors

☐ Hole punch

☐ 4 brass fasteners

☐ String

What You Do:

1. Cut out the back of the cereal box. Lay it flat with the plain gray inside facing up.

2. Before cutting out all the pieces for the Flat Robot, look at the picture on page 17 to see what each of the finished pieces, and the finished robot, should look like.

3. Using the ruler, draw a square on the back of the cereal box with sides that are each 5 inches long and cut it out. Be sure to leave enough space on the back of the cereal box so that you can cut out 4 other smaller pieces too.

4. Draw a rectangle on the back of the cereal box that's 3 inches long and 1½ inches wide and cut that out.

5. Using that rectangle, trace around it to make 3 more rectangles and cut those out too. Now you have 4 cardboard rectangles and 1 big square.

6. With your hole punch, punch a hole near each of the 4 corners of the square piece of cardboard, about ½ inch from the edge.

7. Punch 2 holes centered near one end of each of the 4 rectangles. For the first rectangle, punch one hole ½ inch from the end and the other one ¾ inch from the end. After you punch these holes, place that first rectangle over the other 3 rectangles and use the pencil to mark where to punch the holes in them.

8. Draw a head (or just the letter *H*) for your flat robot centered near the edge on one side of the square. Now you can tell which end is up.

9. Line up the bottom hole that's ¾ inch from the end of one of the rectangles with one of the 4 holes in the square. Place a brass fastener through both holes and bend the ends to attach the rectangle to the square.

10. Do the same thing with the other 3 rectangles. Now your robot has 2 arms and 2 legs.

11. Loop a string through the top holes of the 2 arm rectangles and tie the ends of the string together (see the picture at right). Do the same thing for the 2 leg rectangles.

12. Tie one long string to both the top and bottom loops of string connecting the arms and legs, as shown. Be sure to leave a small length of string hanging down at the bottom.

13. You can decorate and color your robot if you'd like. You can add hands, feet, and other things. Use markers, stickers, more cardboard from the cereal box, brass fasteners, and anything else you can scavenge.

14. To make the robot move, pull on the length of string hanging down at the bottom. The arms and legs of your Flat Robot will move up and down.

What You Need to Know:

A **lever** is another simple machine. It's a support that helps to lift things. For example, a teeter-totter is a fun lever that you can sit on. And even though it doesn't lift anything, the Flat Robot uses levers too.

MAKE A GOBBLE-BOT

This robot moves because of the power of **magnetism.** Even though they may look exactly the same, magnets have a *north* pole and a *south* pole. One magnet's north pole will be pulled toward another magnet's south pole. However, if you place the north poles of two magnets close together, they'll push each other away.

What You Need:

- 2 small magnets
- Popsicle sticks or craft sticks
- White glue
- Duct tape or painter's tape
- A pen or marker
- Googly eyes (optional)

What You Do:

1. Place several Popsicle sticks flat and side by side, until together they're the same width as one of your magnets. It might be 2, 3, or 4 sticks.

2. Take another stick and break off a piece long enough to stretch across the width of the sticks you've placed side by side. About 1 inch from one end of those other sticks, glue the shorter piece across the top of the other sticks. Let it dry.

3. Do the same thing with another group of sticks. You now have the top and bottom jaws of your Gobble-Bot.

4. Now you need to test your magnets to find out which is the north pole and which is the south pole for each one. Hold one magnet in each hand, about 2 or 3 inches away from each other. Do you feel the magnets being *pulled toward* each other? If so, mark the side of one of the magnets facing the other one with an *N* (for north), and mark the other magnet's face with an *S* (for south). Turn both of the magnets over and mark the other sides with the other letters: an *S* on the first magnet, and an *N* on the second one. If you feel the magnets *pushing away* from each other when you hold them up facing each other, then mark both facing sides with an *N,* and then turn them over and mark the other sides with an *S*.

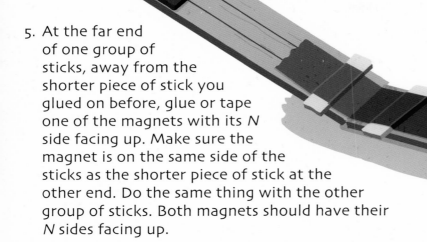

5. At the far end of one group of sticks, away from the shorter piece of stick you glued on before, glue or tape one of the magnets with its *N* side facing up. Make sure the magnet is on the same side of the sticks as the shorter piece of stick at the other end. Do the same thing with the other group of sticks. Both magnets should have their *N* sides facing up.

6. Now you need to connect the top and bottom jaws by making a hinge so that the jaw can swing open and shut a bit. Place the top and bottom jaws together so that the ends with the shorter stick pieces are facing up and are right next to each other. Use lots of tape to cover both shorter stick pieces and hold the two jaws together (just like in the picture above).

7. If you'd like, give your Gobble-Bot some googly eyes. Now try pushing the two jaws together. They should push apart once you let go!

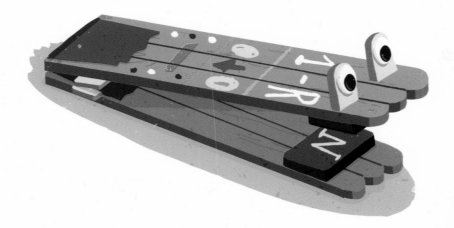

ROBOT DANCER

What You Need:

- 1 cheap battery-operated hand fan (you can find these at dollar stores)
- 3 Popsicle sticks or craft sticks
- Strong tape (like duct tape or electrical tape)
- Some coins (pennies, nickels, or quarters)
- A rubber band

What You Do:

1. Tape one of the Popsicle sticks across the bottom of the hand fan. Then tape another one across the first stick to form an X, like in the picture below.

2. Break the third Popsicle stick in half, then break one of those halves in half again, creating 2 short pieces of stick. Tape those 2 short pieces onto the bottom of the first stick at each end.

3. You should now be able to set your fan on a table. The X will help it stand up.

4. Tape one of the coins to one of the fan's blades.

5. Now turn on the fan. If the fan's On/Off button won't stay on without you holding it down, wrap the rubber band around the fan where the button is a few times until it's tight enough to hold the On button down when you let go.

6. Is your fan dancing? In other words is it shimmying and wobbling back and forth? If yes, enjoy! If not, try taping more coins to the blade. Or if the blades are made of rubber, try bending one of the blades in half and taping it in place that way.

TRY THIS: If your fan falls over when you turn it on, try taping a smaller coin or something that weighs even less than a coin to one of the blades.

ART-BOT

What You Need:

- 3 markers (they should be different colors)
- Scotch tape
- A Styrofoam or light plastic cup
- An electric toothbrush
- Paper
- A table or flat surface
- A box lid or cookie sheet

What You Do:

1. Make sure the caps are on your markers. With the cup right side up, tape the markers to the outside of the cup an equal distance from each other with the caps at the top. The top end of the caps should extend just past the top lip of the cup.

2. Turn the cup over and set it on a flat surface. The cup should be resting evenly on the tops of the 3 marker caps.

3. Now place the electric toothbrush across the bottom of the cup so that the On/Off button is on top. Move the toothbrush back and forth until it seems balanced on top of the cup. Then tape it in place.

4. Tape a piece of paper to a table, or to the inside of a large box lid or the bottom of a cookie sheet to keep your Art-Bot from falling off the edge of the table.

5. Take the caps off the markers.

6. Place the Art-Bot on the paper with the marker tips down.

7. Turn on the toothbrush. The cup should start to move around on the paper, drawing an amazing work of art.

8. If your Art-Bot isn't working, try moving where the markers are taped to the cup. Or try unscrewing the brush part of the toothbrush and removing it. That will make it easier to balance the toothbrush on top of the cup.

ROBO-WOBBLER

What You Need:

- A pull-back toy car
- An adult to help
- Pliers and a screwdriver
- 2 jelly beans, or 2 pencil erasers cut off the ends of pencils
- Duct tape or electrical tape
- Scissors

What You Do:

1. Inside the pull-back toy car is a pull-back motor. We want to get to that motor and use it to build the Robo-Wobbler. If the car is plastic, it will probably be pretty simple to pop off the outside of the car. If there's a little tab somewhere on the bottom edge of the body, pull out on the tab. Or there may be 1 or 2 screws to unscrew. If it's not obvious how to take the car apart, ask an adult to help you.

2. Just 2 of the 4 wheels on the car are connected to the pull-back motor. If you're not sure which these are, try spinning the wheels with your fingers. The ones that spin easily *aren't* connected to the motor, so you don't need them. You can probably pull one of these wheels off its axle, and then slide the axle and the other wheel off the toy car frame.

3. With your screwdriver and an adult, remove the rear wheels and the pull-back motor from the rest of the frame. There will probably be a tab or screw that lets you take it apart. This is what you'll build your Robo-Wobbler from.

4. Cut a 6 inch piece of tape that is exactly as wide as one of the rear wheels.

5. Place one of the jelly beans or erasers against the front side of one of the wheels, near the bottom of the wheel, and wrap the piece of tape around it and the wheel to hold it in place (see the picture to the right). Use scissors to cut the erasers off the ends of the pencils. Instead of jelly beans or pencil erasers, small pieces of cardboard or almost anything else that's about the same size can also work.

6. Do the same thing on the other wheel, but place the jelly bean on the *back* side of this wheel near the top— so it's in exactly the opposite place as the jelly bean on the other wheel.

7. Dress up your Robo-Wobbler by giving it some googly eyes. Or try taping a robot head on it.

8. Now you're ready for action. Pick up the Robo-Wobbler and rotate just *one* of its wheels backward using your hand. Then set it down and watch it wobble!

What You Need to Know:

The motor inside your Robo-Wobbler is a *friction motor*. It stores energy in a spring when you pull the car backward. Other things can store energy too. Think of a rubber band. When you pull back on a rubber band, it's storing energy that's released when you let it go. That's the same way a pull-back car works.

ROBOT BRUSHER

What You Need:

- An adult to help
- An old toothbrush
- Wire cutters
- A pull-back toy car
- Tape
- Pliers and a screwdriver

What You Do:

1. Have an adult cut off the head of the toothbrush. A simple way to do this is to cut it off with wire cutters. Another way to do this is to put it in a vise and saw it off. Throw away the toothbrush handle; we're going to just use the head.

2. Follow steps 1–3 from the Robo-Wobbler activity on page 22 to get the pull-back motor out of the toy car.

3. Tape the toothbrush head upside down to the bottom of the pull-back motor. Now you have a Robot Brusher.

4. Pull your Robot Brusher backward. If you can't pull it backward because the toothbrush head is in the way, pick it up and rotate the wheels backward using your hand. Then set your Robot Brusher down and let it go.

REMOTE-CONTROLLED ROVER

What You Need:

- A remote-controlled car
- A small Styrofoam ball
- Other household items
- Scissors or knife
- Tape
- Glue
- Optional: Tin foil or silver paper

What You Do:

1. Cut the Styrofoam ball in half (or use another item to represent a radar dish) and attach it to the top of the remote-controlled car with tape or glue. This will be your rover that you will use to explore the surface of another planet or a comet.

2. What other items can you add to your rover? Maybe a camera or a drill? Find other items around the house to attach to your rover.

3. If you'd like, you can cover your rover in tin foil or silver paper.

4. Create a rover course in your backyard for your rover to travel over. See if your rover can move over hills and valleys and around obstacles.

TRY THIS: If you want to build other rovers, NASA has some plans online. One is for a Nanorover to explore asteroids (https://spaceplace.nasa.gov/nanorover/en/) and another is for a six-wheel Mars rover (https://opensourcerover.jpl.nasa.gov/#!/home).

FUN FACT: We've sent robots into outer space. For example, in 2004 a spacecraft called *Rosetta* was launched. *Rosetta* traveled away from Earth for 10 years. Then it launched a robot it was carrying to land on a comet!

CODING: SPEAKING THE ROBOT'S LANGUAGE

Imagine a special pencil. You can only use this pencil to *copy* things. The pencil won't work if you try to write your own words with it. You can only copy someone else's words. And you can't draw a picture with it unless you're *tracing* another picture. What a lousy pencil!

That's how many people use computers and robots. They use apps and programs written by someone else. That can be fun. But it's not creative.

So what if *you* could tell a robot what to do? A real robot has a computer inside of it. The computer is the robot's brain. And inside of that computer are many instructions. These instructions are called the computer's code. **Code** is language for computers.

When you write code, you're a **coder**. You are telling the computer what to do. Computers are very good at following instructions. But you have to write *good* instructions. Think about what makes good instructions and the better at coding you will become. For example, if you wanted a robot to feed your dog, you wouldn't just tell it to:

1. Pick up the dog's bowl.
2. Put dog food in the bowl.
3. Set the bowl on the ground.

With these instructions the robot might walk through a wall to get to the bowl. Then it could put just *one* kernel of food in the bowl. And the robot might put the bowl in someone else's front yard. So for this chore the robot's instructions would need many more steps.

Computer code is like a recipe. If you want a robot to make a cake, you have to list all the things you need to make the cake. Then you list all the steps needed to prepare it. Like this:

What You Need:

- 1 cup sugar
- ½ cup butter
- 2 eggs
- 1½ cups flour
- 2 teaspoons baking powder
- ½ cup milk

What You Do:

1. Preheat the oven to 350°F.
2. Mix the sugar and butter together in a bowl.
3. When that mixture is creamy, mix in the eggs.
4. Add the flour and baking powder. Mix well.
5. Stir in the milk until the mixture is smooth.
6. Pour the mixture into a pan.
7. Bake for 35 minutes in the oven.

HUMAN ROBOT: DRAWING

What You Need:

☐ One other person as your partner

☐ A table or desk

☐ Paper and a pencil

What You Do:

1. Sit down at a table or desk with some paper and a pencil. You will be a *robot* and your partner will be a *coder,* who will give you instructions.

2. The coder should think of an object or animal for the robot to draw, but don't tell the robot what it is.

3. Now tell the robot *how* to draw it using clear instructions. For example, "Draw a straight line from left to right" is okay to say. "Draw the roof of a house" isn't.

4. *No cheating.* If the coder wants the robot to draw a dog, for instance, they can't just say, "Draw a nice wolf."

5. The robot can't ask questions. It can only follow the coder's instructions. If the robot makes a mistake, it can only erase something if the coder tells it to.

6. When you finish the picture, see if you can guess the object or animal the coder was thinking of.

HUMAN ROBOT: MOVING

What You Need:

☐ One other person as your partner

☐ A room with a door and a chair or something else to sit on

What You Do:

1. You'll be the robot first. Sit down and don't move until the coder instructs you.

2. The coder's job is to give the robot clear instructions using only 5 commands:

   - *Stand*
   - *Stop*
   - *Turn right*
   - *Turn left*
   - *Go*

3. The coder gives the robot a series of clear instructions to get up and leave the room. The robot will follow each of these instructions so long as they include at least one of the 5 commands listed above. If the coder says something like *Go a little bit* or *Turn left some more,* remember that all the robot can understand are *Go* and *Turn left.*

4. The robot will need to decide exactly what the instructions mean. For example, does *Go* mean to just take one step? Or does it mean to keep going until you run into something?

5. Once the robot has successfully left the room, switch roles with your partner. Try doing this again in your new roles.

GENIUS VERSION: Add a new action for the coder to instruct the robot to do. For example, what if the door to the room is closed? For the robot to open the door *and* leave the room you'll need a lot more commands:

- Stand
- Stop
- Turn right
- Turn left
- Go
- Move right hand forward
- Move right hand up
- Move right hand down
- Move right hand to the right
- Move right hand to the left
- Open right hand
- Close right hand
- Rotate right hand clockwise
- Rotate right hand counter-clockwise
- Move right hand backward
- Push
- Pull

DOUBLE-GENIUS VERSION: For this version, the robot is blindfolded. So in addition to a robot and coder, you'll also need an adult to act as a safety monitor. The adult should make sure there are no dangerous objects or hazards that could hurt someone walking around blindfolded. If the coder sees that the robot is about to run into something, say *Stop*.

LEARNING TO CODE

There are many different types of coding languages, depending on what you want a computer to do. Coding languages have rules, just like any other language. But these can be easy to learn.

One popular coding language is called **Scratch.** Scratch was invented to teach coding to beginners. Instead of *writing* the code, you use building blocks, like the ones in the picture below.

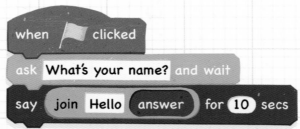

TRY THIS: Visit the free tutorial at https://scratch.mit.edu/tips to learn how to code in Scratch.

Here are a few other popular coding languages:

- **Python** is very popular in **robotics.** It's a good code to learn after you know how to use Scratch.
- **HTML** is a code that is used on most websites. This is what gives a site its look.
- **Java** code is used for Android apps.
- **Swift** is used for iPad and iPhone apps.

FUN FACT: In 1962, NASA launched a rocket into outer space. This rocket had a computer to guide it to Venus. But the rocket exploded! So what happened? There was an overline (like an underline, but *over* a letter) missing from one letter of the rocket's computer code. The rocket cost more than $18 million, so that was the most expensive typo in history.

Do you want to learn more about robots? Lots of schools have their own **robotics clubs** where kids get together to work on projects. If there isn't a robotics club near you, ask your parents if you can visit the First LEGO League Jr. website (https://www.firstinspires.org/robotics/flljr).

There are also many good robot kits out there, from Lego and other companies.

Building robots helps you think like a scientist. It will help you discover challenges in math and science. But more important than that, it's fun!

GLOSSARY

ANDROID (ANN-droyd): A robot that looks and acts like a human. Also called a DROID.

CAPACITOR (cup-PASS-uh-tur): A device used to store electricity.

CODE: A language that provides instructions to a computer.

CODER (CO-dur): Someone who writes computer code using a programming language.

DECISION FLOWCHART (dis-IJ-unn FLOW-chart): A diagram that provides a step-by-step guide for making a decision.

DRONE: A small remote-controlled aircraft that flies without a pilot.

FRICTION (FRICK-shun): The force created by two objects touching or rubbing against each other when one or both of them is moving.

GRAVITY (GRAV-uh-tee): A force that causes objects anywhere in the universe to naturally move toward one other, like making things appear to "fall" toward the surface of Earth.

HTML (aitch-tee-em-ell): A computer code used for designing websites.

INDEPENDENT ROBOT (in-duh-PEN-dunt ROW-bot): A robot that can move and think on its own.

JAVA (JAW-vuh): A computer code used to design Android apps.

LEVER (LEV-urr): A strong bar or other support that helps to lift and move something heavy.

MAGNETISM (MAG-nuh-tis-um): A force that can pull closer or push away objects that have a magnetic material like iron inside them.

MECHANICAL POWER (muh-CAN-ick-ull POW-ur): The rate of work done by a mechanical device, such as the amount of HORSEPOWER generated by an engine.

MOBILE ROBOT (MOW-bull ROW-bot): A robot that can move around.

POWER: Energy used to operate machines, lights, and other devices. The energy can come from various sources, such as water, wind, the sun, or electricity.

PROGRAMMED ROBOT (PRO-grammed ROW-bot): A robot that does the same job over and over.

PULLEY (PULL-lee): A wheel or set of wheels used with a rope or chain to help lift heavy objects.

PYTHON (PIE-thawn): A computer code used in ROBOTICS.

REMOTE-CONTROLLED ROBOT (re-MOAT cun-TROLLED ROW-bot): A robot that's controlled from a distance by a human.

ROBOTICS (row-BOT-icks): The study of how to design, build, and operate robots.

ROBOTICS CLUB: A club where kids work together on fun robot projects.

SCAVENGE (SKAA-vunj): To look for something useful anywhere you can find it.

SCRATCH: A computer code designed for beginners.

SWIFT: A computer code used to design apps for the iPad and iPhone.

TINKER (TING-ker): Someone who likes to build things and figure out how they're put together.